Notes from a Clifftop Apiary

Rowland Molony

Northern Bee Books

Notes from a Clifftop Apiary
© 2014 Rowland Molony

All rights reserved. No part of this publication may be reproduced, stored in a retrieval system, transmitted in any form or by any means electronic, mechanical, including photocopying, recording or otherwise without prior consent of the copyright holders.

ISBN 978-1-908904-60-7

Published by Northern Bee Books, 2014
Scout Bottom Farm
Mytholmroyd
Hebden Bridge
HX7 5JS (UK)

Photography: Stephen Bird, *Devonmovies.com*
Illustrations: *www.EmmaMolony.com*

Design and artwork
D&P Design and Print
Worcestershire

Printed by Lightning Source, UK

NOTES FROM A CLIFFTOP APIARY

ROWLAND MOLONY

From the Journals of a Village Beekeeper

Illustrations by Elizabeth and Emma Molony

Including

15 poems on bees and other creatures

Northern Bee Books

Notes from a Clifftop Apiary

Acknowledgements

Those beekeepers who belong to the British Bee Keepers Association are fortunate in that each month they receive a highly instructive magazine filled with articles about the fine craft of keeping bees. The editor is Sharon Blake. It is due to her forbearance that the articles collected here, which are not at all instructive or scientific, have been allowed to appear alongside seriously worthy pieces which are. To have been given space to be frivolous alongside scientists, researchers and venerable beekeepers of vast experience is a privilege.

Most bee keepers, I suspect, have someone in their lives who 'started them off.' In my case it was Eric Cowen, a man of boundless patience and generosity. I am indebted to him. I am also grateful to Emma and to my wife Elizabeth; both of them produced prints and drawings often at short notice.

RM

Contents

1. Everyone off to the Dance 1
2. Ma Daddy ... He Left Big Shoes 5
3. O M G What a View 9
4. Lobbing Bees by Catapult 13
5. The MI6 Connection 17
6. Ted and Sylvia on the Allotment 21
7. Saving the Greek Economy 25
8. A Christmas Pudding in the Tree 29
9. Chimleys, Queeping and Buddhism 33
10. Marmalade and the Meaning of Life 37
11. A Devon Summer in a Jar 41
12. The Best Way to be Insufferable is 45
13. A Fine Looking Woman with a Beautiful Bottom 49
14. A Vision of the Afterlife 53
15. You Don't Want 500 Mars Bars, Do You? 57
16. Manuel of Beaky Ping 61
17. Hearing Impairment and a Big Moth 65
18. On the Playing Fields of Bulawayo 69

1. *Every one off to the Dance*

A pot of honey. It's a gift. It comes out of nowhere, out of the core-essence of frail temporary flowers. A Summer of sunlight condensed in glass. Molten gold. Can you think of any gift that embodies goodwill, bonhomie, and the simple but profound truth of mankind's interconnectedness with Nature more than honey? Everyone loves to be given a pot of honey. It's worth all the back-break lugging, the sticky mess, the patient filtering and bottling, to be able to give your friends a jar of this year's harvest. I have just finished putting the summer of 2011 into glass pots. I particularly enjoy that final stage in the process. It's a peaceful contemplative act, portioning the distillation of a year's work. I squat on a stool, almost at eye-level with the tank valve, slide open the gate and watch the glassy gold rope of viscous liquid folding down onto itself in jar after jar. It's a present from the year. Don't talk to me when I'm doing this. This takes humility. The old adage is wrong. It isn't more blessed to give than to receive, sometimes it's more blessed to receive, humbly.

From time to time I get together with some musician friends, plus a couple of fellow poets, and we put on a jazz and poetry evening. We've been doing this for some years performing in village halls, churches, basement bars, pubs and on a couple of occasions in a Buddhist Study Centre on the Dart river. Naturally a bee poem gets included at some point in the evening, though I'm beginning to wonder just how many bee-inspired poems can one get, and do they count as yet another 'product of the hive'? Subjects for poems so far have been hive inspections, swarms (naturally), honey-jar filling, and the balmy afternoons of stillness, sunlight and leaf-shadow when absolutely nothing is happening other than the drowsy hum of working bees. Inevitably the swarm poem gets repeated most often, simply because Chris on clarinet does a stunning introduction to Rimsky-Korsakov's Flight of the Bumblebee. He zips through the first ten bars playing fast chromatic scales, till he's at a pitch where he hovers and twiddles and fidgets until your ears tingle and all that anyone can think about is the frenetic tangled whine of insects that's risen to a crescendo …. Then silence. Cue: poem. *Whoa ... Whose idea was this? / Everyone off to the dance...*

There are only so many beekeeping experiences that can inspire a poem, though. The rest is sweating inside a suit in the sun, levering up propolised bee boxes, studying frames and interpreting the story they hold, enduring the exquisite burn of a sting, and re-arranging storage space in the shed. Working in harmony with Nature always sounds great, but most of it is mundane work. I strikes me, though, that there are times in Man's relationship with Nature, when it's Nature that is the dominant partner. Take four-leaf clovers; take honey. You don't find four-leaf clovers, they find you. You'll waste hours of eye-on-the-ground time searching; but one time when you're not looking, your eye falls on one. And then honey. You don't eat honey. It simply sinks into you. Put a spoon of it into your mouth: your body absorbs it, you don't even have to swallow, natural sugars spread out and down into the fabric of your being, leaving you with a smile on the inside of your face.

❦

So now we are up at the allotment, Wife and I. We are clearing away overgrown scrub and tree cover to give the bees ventilation and maximum sunlight. Autumn is the start of the beekeeper's year-cycle. Everything we do now is groundwork prep for Spring: so we ensure they have a secure hive that excludes mice, woodpeckers and livestock, sufficient food for five months, a fertile queen, and a healthy population of over-wintering bees. The overhanging branches are coming off because nobody wants the unnecessary disturbance of raindrops pattering onto the roofs. And I'm thinking too of those windless January days when a wintery but warm sun lights on the hive and the bees come out to empty their little bowels and then linger to soak up the sun. We want all the warm light that's available to

fall on them, unshadowed by trees. The hives are on open mesh floors, but the crown boards are being lifted very slightly and feed holes covered to encourage an even and circular air flow and to avoid the funnel effect.
We take a break for a moment from the hauling of hazel and blackthorn branches.

"What label are you using for your honey pots next year?" Wife asks.

"The usual," I say. "The photo of the hives with the sea-horizon in the corner."

"I think it's time you had a new one," she says. "Someone in a bee suit maybe, holding up a jar of honey -- with a label on it showing someone in a bee suit holding up a jar... etcetera. What about that?"

"But people know my label," I say. "They're familiar with it. It's a brand label. Besides, I don't really want to be pictured on my own honey pot."

"I wasn't thinking of you," she says.

Ah.

"What if I had your bee suit on and I stood like this?" She alters the pose. "Or how about this?"

I pause. Much might depend on my answer. "I'll think about it. It's an interesting idea. Modelling honey."

"I think you could shift loads more jars."

I make a mental note to order more of my present labels. Lots more.

Notes from a Clifftop Apiary

2. "Ma Daddy ... he left big shoes..."

Through the gate, down the path, past all the other neat and well-tended allotments... to the tangle of bramble and the waist-high grasses that hides the hives. I am lucky to have this hidden corner plot. Behind it, beyond the blackthorn hedge and filling half the horizon is the sea; today it is calm and glittering and streaked with meandering tracks, like paths across a plain to the taut skyline. The moods of the sea.... The mood of the bee keeper ... The mood of the bees ... Life is infinitely variable. Today I am willing to open up the hive that has defensive bees to have yet another look for the queen, to mark her. It is not always so; some days I start the job, but after a few minutes I give up, put things back together and move off. Some days you just don't feel like doing it. Today turns out to be one of those days. I can't find the queen and the bees are irritated. I re-assemble the hive but keep the suit on to pick blackberries. It's been a very good year for blackberries. But the bees have been roused now and it's difficult.

Frank the allotment neighbour leans on his hoe and looks across at me.
"Gardnin' in your bee suit ... must be hot?"
"Bees a bit cross," I call back.
"Well, you shouldn't rile 'em. Leave 'em alone to get on by their selves. That's what they bio-dynamicals do."

This is a revelation, Frank knows about Biodynamics? I'm tempted to take it up with him, but his back is bent and he is deep among the rhubarb. It wouldn't be surprising if Frank does turn out to be sympathetic to the Biodynamic movement because he is a master vegetable grower. I doubt, though, if he sows by phases of the moon, stirs his water butt with a cow's horn, or if he involves the cosmos in decisions about mixed plantings. He just has that touch, and he pays attention to detail.

In the field of bee-farming, having the touch, and going in for detail can mark you out as a certain kind of keeper. The Biodynamic folk tend to go for top bar hives, and favour minimal management. They also 'honour the sanctity of the swarm'. This might be fine if it was allied to up-to-the-minute knowledge of disease pathogens, viruses and current research in integrated pest management, but one reads little of this in the literature. Arguments persist over 'standard' bee keeping practice as encapsulated in the BBKA's examinations, and minimal management practices represented by believers in the Warre hive. But the end of the argument always comes to rest on this fact: your bee keeping practice is a reflection of you, your temperament, your sympathies, your capacity for detail and fine-tuning. Conclusion: if it works for you, that's the way you'll choose to do it. Go intensive bee-farming. Or let them build their own comb, add space from below, take a % of honey but leave enough not to have to feed for winter. Nobody is correct and nobody is wrong. Like politics and religion: your choices reflect you.

I'm a convert to propolis. In the shed I fill up little phials and bottles and jars with it. This comes home with me.

❦

I make it a regular practice these days to gather and keep propolis. It takes patience to scrape it off and collect it. You can do this by dropping the crumbs onto a white sheet, or into a plastic bowl. In its natural state propolis hardens to beads of crystalline resin. More often, though, you'll come across it in the hive sticking everything down and making the whole living box tight and secure and draft-free. It's mixed with saliva and wax to make it pliable, so that the bees can mould it to fit gaps. Clearly, bees got around to grouting long before we ever thought of it. The biggest gap of all in their home is of course the hive's front door. To assist the guard bees, they will reduce the size of the entrance by using waxy pliable propolis to make a curtain or short wall of 'propwax'. (*Pro polis* means, in the forefront of the city.)

The marvellous thing about it, and what makes it so collectible, is its medicinal properties. If you suffer from a sore throat, try dissolving it in brandy and using it as a gargle. It has an anaesthetic effect; it's also antiseptic. Some people take it at the onset of a cold. Propolis taken from the hive is composed of balsams and resins, waxes, oils and pollen. Tree buds and barks secrete it to protect against moulds and infections. Bees use it to strengthen comb at the point of attachment, they polish cells with it, and they embalm dead intruders, like mice. Some people chew it. This is okay if you're staying at home, but don't do it if you're going out among ordinary human beings, and definitely not if you're going to a party, because you'll have bits of bright orange stuck in your gums, and people will freeze when you smile because your teeth will be bright yellow.

There has been a brief correspondence in the BBKA News about breathing onto bees to make them scatter from tight clustering. You have to feel sorry for them. It's bad enough for you or me to have someone with bad breath talking into our face; how much worse if you're a bee if a keeper with halitosis leans down and breathes on you. It's not funny. You'd scamper off that frame as fast as your little insect legs could carry you. The patron saint of bee keeping, St Ambrose, got the job because it is said that when he was in his cradle a swarm of bees settled on his mouth. Maybe he attracted them with his milky baby breath. There is no record of whether or not the other Father Figure of British bee keeping, Brother Adam, had sweet breath. Both were churchmen. Ambrose (340 to 397) was Bishop of Milan and Brother Adam was a German monk based at Buckfast Abbey. Maybe the breath of God passed through them. The Biodynamicals would approve of that.

It's Friday night and we are in the pub drinking to the memory of a deceased beekeeper. Sitting with us is the deceased's son, over from America for the funeral. Conversation is stilted. In the end I ask the son if he, like his father, ever kept bees. There is a long pause; son tightens his mouth, squares his shoulders and looks away. "Naw," he says eventually. "Ah din' do that. See, ma Daddy ... he left big shoes." The company narrow their eyes and nod. "An' ma Daddy, he cast a long shadda." The company nod again and take a pull on their pints. Big shoes, and a long shadow. I get some more pints in. It's a long night. In the morning Wife gives me an odd look. I ask her what's the matter? She says, "When you got in last night you told me that when you die you're going to leave me your big pants, and all your records by the Shadows."

3. O M G What a View

Sliding the white board inspection tray out during the summer months I usually find, among other debris, small lumps of pollen which have fallen through the mesh floor. Collecting a few and putting them on my tongue produces a small powdery explosion of nutty flavours. This year I saw white pollen being carried up a landing board on 16th January. This would almost certainly have been pollen from the male hazel, *Salicacea*, which is always the first to put out. At the other end of the flowering season, October and November, it's the pale orange of ivy. These, the first and the last pollen producers in the year, are easy to identify. Right now though, in May, the whole complex colour-range of summer pollens are arriving.

A few days after the white pollen sighting, 23rd January to be precise, -- a windless day of clear blue-washed skies, with blinding sunlight flashing off a shining sea -- the air felt mild enough to take the lids off the hives and do a quick check on stores. I don't like hefting. It's a very imprecise measuring practice, it usually means cracking sections like brood box and floor apart from each other. Sometimes it's brood box from super and, because you don't have a hand free to brush bees away, this always results in crushing those bees that immediately crawl into the opening. And can I tell what the poundage of stored food is in there, separate from the weight of the boxes? No, I can't. So, (another benefit of living in South Devon) opening up on a mild midday in January does no harm, and means I can quickly spot frames holding late-summer honey which the bees made after I had taken off the summer crop in early August. So that was alright then. It was yet another reminder that rural Devon, and the coastal margin, is a very fruitful region in which to keep bees. I could, if I wished, take two or even three honey crops during the course of a summer. As it is, I take honey off in August, and whatever the bees make after that they keep for winter. Which reduces the syrup feed.

※

There's a wooden five-bar gate at the roadside which leads to the village allotments. Whenever I take visitors to see the bee hives, I hear a sharp intake of breath at the gate. "O M G, what a view!" Well, yes. I spend a lot of time looking at it. A ruler-straight horizon. The wedge of Portland. The distant hills across the bay above Abbotsbury. And the infinite faces of the sea. I'm the only vertical gardener there. All the other allotmenteers are bent backs. They're rooting among the rhubarb, fossicking among the flowerpots. I'm the only one standing. I watch waves travelling inshore, like curved ribs. I can hear the slap and rinse of small waves on the shingle beach far below. On vast blue windless days I stare out into sea-glitter. This is not time wasted. When I gave up full time work, my colleagues looked hard at me and said, "But what are you going to do?" They, like me, had spent half a lifetime with their mind in next week, next month, and found it difficult to imagine a daily life drained of future-content. A life of conscious attention in the present moment. But that's all in the past, and here we are, it's Now time. And keeping bees only contributes to that. Zip the suit. Light the smoker. Select the tools. Pause by the hive and watch the bees. Then quietly open up the box, and quietly look in. Everything done consciously in Now. Deliberations rooted in the present. Quality time.

※

I am on the allotment and looking at brood frames when my mobile chirrups. A text from Wife. *Whoosh u doomigingi?*

I ring her. What is her message about, I ask? She says she is still trying to get the hang of predictive text on her new mobile. She wanted to ask, *What are you doing?* "But," she adds, "I don't want you to ring me, I want to do texting"

"Why?"

"It's more fun."

"But it's quite difficult to operate a mobile phone from inside a bee suit and wearing marigolds."

"Well, I'd come home now anyway if I were you because your lunch will be cold."

When I get home I find lunch is quiche and salad. "I thought you said I had to get home because of lunch getting cold."

"Quite right. There it is. It's cold. It's a cold lunch. Are you complaining?"

I tell her I think I'll leave my mobile at home in future.

"No, you can't do that. You have to have it so you can call the ambulance."

"Call the ambulance? Why?"

"For when the bees attack you and you have one of those naff-electric shocks."

Notes from a Clifftop Apiary

4. Lobbing Bees by Catapult

Each year my fellow allotmenteers get a free pot of honey. It's by way of compensation because maybe in July a ball of bees appeared inside a rhubarb forcer, or back in May a giant's beard of bees arrived to hang off somebody's bean canes. They are, as are all human beings, a colourful and mildly strange bunch of people. Sometime in the Summer a notice appeared by the entrance gate: "Kindly refrain from allowing your dog to commit befoulagement of the environs."

"Seen my notice, then," said neighbouring plot-gardener Frank.

"It's impressive," I said. "It's got style."

"Ah. Not afraid of long words, me. You take your Dickens and Trollope and your Henry James. Them fellers, they didn't dumb down. They didn't stoop."

"No. Right. So, your notice about dog poo is up there with Dickens and Trollope, is it, Frank?"

"Been goin' to this creative writing class, see." He straightens, leans on his hoe and screws his eyes to look across at me. "You does something along them lines, in a small way, don't you?"

"In a small way," I agree.

Frank tells me he's started a novel. He's calling it A Right Old Carry On Up the Allotment. "And it en't got no sex in-you-endos about vegetables in it, neither." He pauses and gives me a meaningful look. "I might bring your bees into it."

"Okay. Anything you want to know, just ask."

"Ah, and if you need me to check over your stuff before you sends it in, you just say."

Rainfall records have been kept in East Devon since 1860. In the intervening 150 years, the only summer wetter than 2012 was in 1876. The district's honey yield for the summer reflects that. Through luck and location, I managed to spin off enough for family, friends and my landlord. Why give some to the landlord? Because there's something weirdly English and historic and traditional about passing a pot of honey up the line to the hereditary land-owning peer of the realm who happens to own my allotment as well as a hefty chunk of the rest of Devon as well. My annual rent probably keeps him in cigars for a week.

Do bees benefit from living beside the sea? I think they do. The sea moderates extreme weathers. On-shore winds are frequently balmy. Severe temperatures rarely last long. And the light is a factor. The warmth of early sunlight on the hives invites an early start to the working day. And frequently sunlight is intensified by being reflected off the sea. The sea and its infinite faces: one day it's flat as metal under sheets of blinding light; another day it's a theatre stage spot-lit from holes in the cloud-ceiling by slanting shafts of sun. All this has a very beneficial effect on the beekeeper too. There is often a quality of stillness in the early windless air before the day has got under way. Stillness in Nature is infectious. When you notice it, how nothing moves, how every leaf is utterly motionless, then you join it. It's called Meditation. Dusk often has this special quality too. The hives are lumps of shadow, humming quietly like buried motors. Every leaf, every twig in the gloaming is perfectly still. Forty miles away over the dark waters of the bay pin-prick lights flash for a moment on the Abbotsbury Road and then go out. It's where cars crest the rise and then dip again above the long shingle bar of Chesil beach.

"Did you know that bees protected sacred sites in history?" says Wife, who has been reading about such things. "They protected the special religious places of the African Bushmen, otherwise known as San. And the Minoans kept bees. And the Ancient Egyptians." She looks at me meditatively. "You have bees. Maybe you're a Shaman."

"No, thank you, I don't want to be a Shaman. I'm quite happy being me."

"Mind you," she adds, "the Romans used to lob hives of bees by catapult at their enemies. So not all bee keeping people are civilised."

I make a note to tell Frank about the lobbing. He might put that in his novel.

5. *The MI6 Connection*

Those torrential late-November and December rains of 2012 that flooded the river valleys of south Devon stained the inshore waters of Lyme Bay a thick mud-red. Waves looked hacked and chiselled under the south-west winds, and all the way out to the horizon the sea reflected a brilliant hard light. Plodding up the hill to the allotment was like wrestling with invisible forces. Under the battered blackthorn hedge the hives looked vulnerable and beleaguered. There were rows of rain drops hanging under the roofs and damp stains streaked the brood boxes. Later in the dusk when the wind had dropped I went back up there; found a vast quiet after the tumult. Nothing moved. Every leaf, every twig, perfectly still. The hives murmuring. Far out in the darkness of the sea there were lights of ships separated by miles from each other, twinkles on the horizon signalling a particular kind of loneliness. This seemed a special sort of end to an unruly day. After the rough times come moments when the keeping of bees on the top of a cliff above a southern sea becomes a privilege.

Honey bees are studied quite extensively at Princeton University. James and Carol Gould are researchers there and they have published a book called **Nature's Compass**. It's about the astounding capacity of wild creatures to navigate their way over the surface of the globe. Curlews fly 6,000 miles non-stop along invisible bird-flying lanes in the sky as they travel from the South Pacific to Alaska. Spiny lobsters crawl in an antennae-to-tail conga for 30 miles along the ocean floor. Idaho salmon travel 900 miles and ascend 7,000 feet in elevation seeking and finding the tiny creek in which they hatched. And bees ... Bees apparently don't like to fly over water. In one experiment a food source was placed in a boat, forager bees having been fed on the source and released. At the hive the dance was observed to be vigorous, but the observing bees didn't pursue the food. The authors' interpretation is that the bees decoded the distance and directional cues, placed the location on their mental maps, understood that it was in the middle of an adjacent lake and refused to act on the information. When the site being signalled was on or near the farther shore, then the bees were willing to fly to it. Perhaps, then, forager bees retain a mental memory map of the areas surrounding their hive, in the same way that we carry in our head a map of all the land features which surround our own house.

Not long ago a full page advertisement for MI6 appeared in our Sunday paper. This caught the eye of a certain person in this household. I catch her glancing across at me, then back to the paper, then me again.

"I think you should apply," she says.

"You think I should join MI6?"

"You're just right. They're looking for ordinary. They don't need hyper-intelligent. And they want people who don't talk much."

"And what would I be spying on here in Devon?"

"You're always looking out to sea. You can keep watch. For enemies. And your cover is perfect. Beekeeper. Disguised in a suit, head-to-toe."

"But exactly who would I be---"

"And if you have to follow a suspect inland you can take some of your bees with you in a matchbox and attach micro-messages to their legs, like they did with pigeons in the first world war, and they'll home-in on your clifftop headquarters. Perfect again!" She smiles, obviously pleased with her idea.

I sit back and close my eyes, realising yet again why it is I spend so much of my time up at the allotment, with just the murmurs of bees, the cries of gulls and the distant crunch of the shingle.

"You realise," I say, "If I'm in MI6, I'll not be coming home from the allotment and telling you all about what I've been up to?"

"Really?" she says. "Surprise me with something new."

※

It's not often Wife accompanies me on visits to the Branch Apiary. On one particular day I had a specific errand.

"This is a bumpy lane," she remarks, "Where are you taking me?"

"To the Mating Apiary."

"Really? Oh." There is a silence while she absorbs this. "Well," she remarks finally, "you're a dark horse. You certainly know how to show a girl a good time."

Not being sure she has understood the set-up, I emphasise, "It's the *Branch Mating Apiary*."

She is silent for a moment. "I know. You just said that. Do all you beekeepers come down this lane? With your wives?" I am about to try to explain when she says, "Are you going to tell me what will happen there, or are you leaving it to my imagination?"

I'm still not sure to this day if she was disappointed or relieved at the way things turned out.

6 Ted and Sylvia on the Allotment

Sometimes we have friends up on the allotment for a lateish sort of breakfast. Cook up a bacon bap, and brew mugs of tea with a kelly kettle. Or if it's later in the day, take scones and jam and cream up there. The bees are fine with that, so long as we don't sit in their flight line. Some folk, though, are understandably nervous. It's not every picnic site that your view out over the sunlit sea is criss-crossed continuously by zipping bees. They don't bother us. Some visitors were amused that I'd painted red, white and blue RAF roundels on my landing boards. Well, why not? I have an affinity with the RAF. My father flew Bristol Fighters, Sopwiths and S.E.5s in the RFC in 1918, and he commanded a Barrage Balloon Squadron in the Second. And he put me into the RAF Boy Entrants in the early 60s. What else are bees if not pilots, navigators, flight engineers, and in early spring, bomb-aimers?

There are quite a lot of people I would like to have invited onto the plot to look at the bees and look at the view and chat about whatever these things gave rise to. Writers, mostly. Like Sylvia and Ted. The arrival in the post of the County Yearbook of Beekeepers makes me think of them. How, if things had turned out differently, they might have become successful beekeepers, perhaps have appeared in the Okehampton Branch list. Would she have kept her own name on her honey jar labels? What price today a jar of Sylvia Plath's honey? Only, she never got around to keeping bees through a full year-cycle. She wrote a small group of poems about bees. Ted didn't. Somehow, bees didn't get his creative juices going.

Some years ago I wrote briefly about Sylvia Plath's bee poems in the Devon Beekeeping Magazine, edited by Glyn Davies. This came to the attention of the Keeper of the Plath Collection at Smith College Library in Massachusetts, Plath's Alma Mater, who kindly sent us photocopies of pages from Sylvia's journals from the Autumn of 1962. That summer she and Ted had met Charlie Pollard and a group of North Tawton beekeepers, and by October she had received delivery of a box of bees. A sequence of five bee poems were written intensively one after the other in a matter of a few days in early October. Of these "The Arrival of the Bee Box" is probably the most accessible and therefore best known. Its mystery and potent danger set her off:

The box is locked, it is dangerous.
I have to live with it overnight
And I can't keep away from it.

The excitement of fear animates her writing:
I put my eye to the grid.
It is dark, dark,
With the swarmy feeling of African hands...

How can I let them out?

In her writing about bees, she never developed beyond a sense of mystery tinged with fear. Did she ever get to the point of having a hive in her garden at Court Green? If she had become familiar with an occupied hive and with handling bees, then she might have written more genially about them. But there was much else that filled her mind at this time. She had two young children. She was driven by a powerful writing ambition. And her marriage was coming apart. Ted by this time had published two volumes of poetry[1]. Among them a handful of 'nature' poems of stunning power and originality.[2] It was a household containing two artistic literary ambitions. By the end of the Summer of 1962 they had parted. The local midwife who had attended Sylvia at the birth of her son, Nicholas in January, Winifred Davies, was also the woman who taught her beekeeping. Elsewhere among the bee poems Sylvia makes much of the fecundity, power and authority of the queen. Some commentators see in this a resolve in the writer to re-form herself as an independent woman in control of her own creativity. Obviously, after the breakdown of the marriage, she had no taste for staying on in the little Devon town on the northern rim of Dartmoor. She took her two children and left North Tawton for London. There were further poems to come, terrible poems that steered inward towards darkness and annihilation; certainly nothing more about bees. And rest, as we know, is silence.

So, yes, Ted and Sylvia up on the allotment for a brew-up and look into the hives. Why not? It's an attractive whimsy. Playing the "what-if?" game. Those of us who survive the dark periods of life and go on to have this working partnership with bees, flowers, pollen, nectar and honey --- well, it's a pretty life-affirming occupation, isn't it? One could wish it like a beneficence on others who might have found in beekeeping a living lifeline, a connection with the core of nature. And be continually reminded of it by the zipping of flying bees coming and going across the forefront of your life.

---

[1] *The Hawk in the Rain* (Faber 1957). *Lupercal* (Faber 1960)

[2] The curious reader who may be unfamiliar with the volumes might be rewarded by looking at The Thought Fox, The Horses, Wind from *Hawk in the Rain*; and to Esther's Tomcat, Hawk Roosting, View of a Pig, November, Thrushes from *Lupercal*

7. Saving the Greek Economy

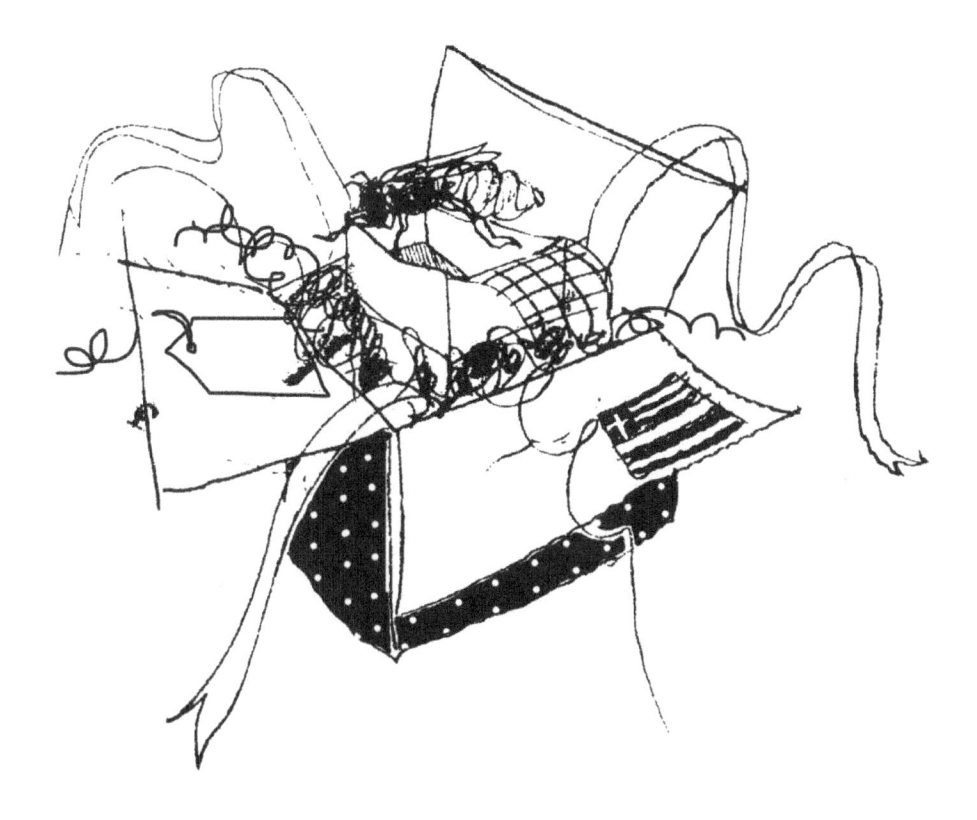

How's this for the loosest, most tenuous connection between beekeeping and the world of literature? Or, Literature. The small wood and canvas folding stool that I sit on when I'm operating the valve on the honey storage tank and filling the jars once belonged to Gilbert Miller, producer and theatre impresario, who produced the plays of T S Eliot in New York and who owned The St James Theatre in London. It's not a flash stool, not like a Director's chair with his name printed on it. In fact, I wonder just when he would have used it. Maybe he took breaks from rehearsals by squatting on his little folding stool and chatting to Eliot. Maybe he offered up the stool to Eliot? Who knows where historically significant bums have parked themselves? I once sat on the ancient settle in the fireplace of Ann Hathaway's cottage. There's not much of it left now, Americans having picked off splinters for decades and taken them back across the Atlantic in order to regale their friends at dinner parties with the story. "Have you seen this? See that splinter of wood? Shakespeare's ass rested on that. No kidding. Ever seen anything like that before in your life?"

I remember a particular dinner party on this side of the pond which was memorable not for any connection with Shakespeare's bottom but with a bee sting. Sometimes a sting feels deserved, because you were clumsy. But sometimes it just comes out of the blue. There I was, standing between a couple of hives looking out to sea, admiring the slow ridged waves sweeping powerfully across the bay, when Zap. Right between the eyes. I scraped the barb out, but the venom was in. I felt I'd faced a firing squad, been shot square in the forehead, but survived to feel the pain.

"What's the matter with you?" Wife said when I got home. "You look like you've got something on your mind."

"I got stung."

"You remind me of somebody," she said, studying me. "Who is it...? Oh, I know, that actor who played Frankenstein's monster. What was his name?"

"Boris Karloff."

"That's the one. You look like him. Why are you crossing your eyes?"

"I'm not crossing my eyes. Why would I do that?"

"Well, it looks as if you are. Anyway, have you forgotten we're going out to dinner tonight?"

"Oh?"

"Are you going looking like that?"

"Is it fancy dress? I could go as Boris Karloff?"

"No, you can't. If you turn up looking like that they'll have to keep their children locked in their bedrooms."

In the end I wore my fishing hat with the brim down and told people I was protecting a wound. After several bottles of wine the company, the wives mostly, demanded that I show them my wound.

"It's a bee sting, actually," I said, lifting my hat off. Suddenly the jollity stopped. There was an uncomfortable silence. Somebody coughed. Wife said in a small quiet voice, "I think you'd better put your hat back on."

We left shortly after.

Swarms came late this year. I did in the end collect two, both had lodged in garden hedges in the village. One in a fellow beekeeper's hedge, and the other in a fellow allotmenteer's hedge. This, in a village with many hundreds of gardens, how uncanny is that? *C'est mysterieux.* Right now, in November, the little blighters are busy hauling ivy pollen and nectar back into the hives. Ivy is one rare instance when the beekeeper knows better than the bees. Ivy is useless. I first heard from a couple of old Devon yeomen what ivy was good for. "You'm don't want none of that stuff," they told me, " 'tis a right b****r for Granny Lies In." Later the penny dropped. Granule-lizing. Ivy stored in the cells becomes hard and useless. And if you take it off and jar it, it still tastes unpleasant. So, all round, a bit of a curse.

❦

"When is your birthday?" Wife asks.

I tell her. Each year I have to tell her. I never have to ask when hers is.

"I've got you something for your bees."

"Oh?" I am cautious about this, remembering the time I was given a cardboard box filled with the shredded paper packing that protects consignments in the post. I rootled around in it for some time before Wife said, "What are you doing? There isn't anything else in there. That's it. That's the present. It's shredding. For your smoker."

"So ... this present. Is it made of stainless steel?" I was picturing just possibly a motor-driven eight-frame radial extractor.

"No. It's a new queen. From Greece. A Greek queen. Doesn't that sound marvellous? She's flying over from Athens. Not on her own. In an aeroplane."

"A queen? From Greece?"

"She's been bred on the mountain slopes where the parsley, sage, rosemary and thyme grow."

I tell her there is a debate about the wisdom of importing queens from abroad. "Some people think it's not a good idea."

"Look," she says, "these are hard times for the Greeks. Their economy is going down the toilet. I don't suppose you've noticed, but I've been buying Cretan olive oil from the Co-op."

"And this queen that you've bought is going to save the Greek economy?"

"We all have to do our bit to help." She looks thoughtful. "This queen's already mated. So your drones won't have to do anything. I feel sorry for queens. A lifetime of pushing out babies, and she's only ever had sex on one or two afternoons in her whole life. But then it was with the whole rugby team, so maybe it was pretty memorable."

8. A Christmas Pudding in the Tree

Give bees a disused chimney, or a fissure in a cliff face, or a deep space in a hollow tree and they will build comb up to, or rather down to, a couple of metres in length. And if in their natural unmanaged state they're prepared to do that, why stint on giving them lots of space in the Spring? I've never seen any point in delay about giving bees room for expansion. Lots of it. Brood space for the nest, and storage space above for the honey. And in plenty of time. Give them elbow room. Head room. Dormitories. Lumber rooms. Raise the roof, literally, so the impulse to swarm gets lost in the rafters. There is a drawback, though. Big honey storage boxes require a strong back. A brood box filled with honey frames can weigh well over twenty kilograms. Lifting that could do you damage. In the past I've had to take out the frames individually and half-empty the top box before I could lift the whole thing away. Commercial beekeepers whose management strategy aims for around 150 pounds, or 75-plus Kgs, per hive, and who calculate their total harvest in tonnages, will have their own removal system. It's probably called two-men on the job. The hobbyist beekeeper, however, may well have to call on the help of a partner in taking off the honey preparatory to extraction. This can bring its own difficulties:

"Do you have to stand on that chair? It's not safe. If you fall off I won't be able to pick you up, not if you've got a box full of honey on top of you ... I'm getting very sticky here.... There's a bee's bottom sticking out of one of these hexagonicals, does that matter?... Aren't people going to complain if they find bits of bee stuck to their teeth?... Are all these drips going into my kitchen?... Who's going to be cleaning the kitchen floor I should like to know ..."

Some parts of the extraction process can be enjoyable. Lining up the clean jars within arm's reach. Sitting on the stool and opening the valve on the settling tank. Watching the glossy tongue of honey curving down into the jar, watching the honey rising to the top like a sunny smile to the glass lip. Bounty. Riches. Freely given from Nature. This is a form of richness that has no connection with economics.

Another gift freely given is the swarm. Being called to come and collect one is usually a pleasure. But it helps if the phone call comes to me rather than Wife. I came into the house one late morning to find her on the phone.

"It's where? It's hanging in a tree. Well, keep your pets and your children away from it. How big is it? Football? Rugby ball? A pudding. What sort of a pudding? A steam pudding? With raisins in it? Lots of raisins. And currants. Currants and raisins. That would make it like a Christmas pudding. Would you say it was like a Christmas pudding? A Christmas pudding in the tree. I see. And how many would you say there were? No, you don't have to go out and count them, just a rough ... what? Oh ..." She hands the phone to me. "I think she's gone."

I didn't get a swarm this year. Given the prices currently being charged for a five frame nuc, a swarm suddenly becomes a pretty valuable item. Capturing and retaining your own now becomes important. To that end it's worth having a bait hive. Each Spring and Summer I leave a bait hive out on the plot. Several times over recent years I have come to the allotment and found a previously empty hive now suddenly full of bees. Presumably, one of my hives had swarmed and immediately found a new home just two doors down. Very convenient for all concerned.

❧❧❧

The wildlife has kept us watchful on the allotment this summer. We have a family of foxes. I see the faces of the cubs peering at me from behind the rain storage butts. A pheasant has been enjoying dust baths in the cold ashes of my bonfires. For a period, until it got fed up with being disturbed by my presence, a quite large adder used to sun itself between my courgette plants. I've watched a pair of peregrine falcons circling high overhead. In my corner of the allotments there are green finches and blackcaps, fly catchers and tree creepers. And a robin who'll take cake crumbs off my palm. Just as his dad did last year. Maybe it still is the dad. How can you tell robins apart? Only when there are two of them on the bird table, one feeding the other. The presence of bees doesn't in itself attract wildlife. Although I'm told Guinea fowl are not a good idea near an apiary. A guinea fowl has been known to stand in front of the hive entrance and pick off every bee that showed himself on the landing board until he was full. There's also been a dolphin in the bay. I've spotted his gunmetal grey back arching through the waves, the spume from his sighs flying away in the breeze. A bit of a media tart, he's called George. He turns up each summer and does the season, gets his picture in the papers, and performs a daily dally with the swimmers just off-shore.

Every now and again Wife has to make her feelings known about the clifftop apiary.

"I'm a bee widow. Did you know that?"

I ask her what she means. She says, "A bee widow, you know, like a golf widow. I'm one of those women whose husband is always off out and away – in your case, with bees."

"Isn't that better than being in the pub all the time? It could be worse. I could be into Amateur Dramatics. Or I could be a fitness fanatic."

She gives me a cool stare. "Or you could be one of those husbands who have a really big tool box and they're always putting up shelves and building cupboards and fitting new kitchens and bathrooms."

"You want me to fit a new bathroom?"

"Absolutely not," she says.

9. Chimleys, Queeping and Buddhism

A Day in Summer.

a.m.

It's a breathless morning up on the clifftop. The sun across the bay is gold on sheet-metal, the yachts on the horizon are going nowhere. Nothing is moving – except bees. And Frank. The bees are paying floating visits to bramble blossom and waltzing down to their landing boards. And Frank is hoeing his rows. Frank is my allotment neighbour.

"Hot in they hives," he calls.

"They're not all inside, Frank. Some are visiting your beans."

"Too hot," Frank persists, "to be shut up in one of they boxes. I wouldn't want to be stuck in one."

I look at my hives. One is in the shade of a hazel tree, two are partly shaded by blackthorn. One is in full sun.

Frank leans on his hoe. "Ever wonder why they likes chimleys? When they takes off and swarms, they likes to lodge in chimleys 'cos they got a flue. For the breeze, see. Bees likes ventilation. You got some air-flow through they boxes?"

Frank has a point. Frank always has a point, dammit. I take a look in my most heavily populated hive and find beads of condensation on the underside of the roof. Off comes the crown board. In other hives I insert pieces of wood between the crown board and the roof to assist air flow. Who knows if this is necessary? Sometimes one has to go with an instinctively good idea. There'll be no harm on a hot day, at least. This is not the first time Frank has hit a topical spot in the matter of bees.

"Frank," I call, "why don't you keep bees?"

He pauses in his hoeing and looks at me from under bushed eyebrows. "Bees comes in thousands, don't they?"

"Yes"

"And they's all female."

"Yes"

"I got just the one of them at home. That's enough."

I read up on ventilation in the beekeeper's bible: Wedmore's Manual of Beekeeping. He says that bees will not like a powerful gale blowing up their passage. Quite. Who would? He goes on: *Stocks cannot be killed by cold air alone so long as it is only in slow motion in the hive.* A gentle air flow is desirable then. What else? *Evaporation of nectar in the hive, and the removal of the vapour by fanning bees, absorbs large quantities of heat at a time when the outside temperature is high.* Brilliant, they don't really need me at all, do they? But I ought to provide water nearby.

Midday

Back home, Wife does not agree with Frank or with Wedmore. "Your bees are mostly girls, they won't like a draught of any sort anywhere." She is looking at photographs in a book on bee keeping in the tropics. "I want one of these," she says. She points to a photograph of what appears to be a small coffin dangling from the upper branches of an acacia tree. "It's a hollowed-out log. Hung out of reach of predators. See, they've wired it together, and there's a hole for the bees to build wild comb inside it. Now isn't that much more interesting than your boring boxes? Come on, isn't it?"

I study the picture. "You want a hollow log? On the allotment?"

"Yes. Up a tree. A hollow log really resonates with me."

"Where exactly were you thinking of putting it?"

"Sling it from a high branch off one of those wild cherry trees around your shed. Well away from the predators."

"There aren't any predators on my allotment."

"Oh, come on. There must be some, surely?"

"I saw a rabbit there once."

"Anything else? They have honey badgers in Africa. What about wild boar?"

"There's a mouse in my shed."

She is pinching the bridge of her nose and her forehead looks creased. "Anything else?"

"No"

"Look, I'm going to all this trouble of hollowing out a log and wiring it to a high branch to make a really attractive tropical bee hive, so the least you could do is come up with some predators."

Lunch is taken in a thoughtful silence.

pm

A balmy afternoon. I am sitting outside my shed in the shadows of the wild cherry trees nailing up frames and slotting in sheets of fine smelling wax foundation. It takes twice the time to recycle an old frame – scraping wax out of every slot – than it does to assemble a new frame. But I don't mind. This is calming and therapeutic. Overhead there is the unhurried hum of bees zipping and unzipping pockets of the afternoon: the wild cherry trees are in flower. Then it's time to check one of the hives for queen cells. The bees are preoccupied and take little notice of me. Most frames are clear of queen cells, but one near the centre of the nest makes me hesitate. What is it? I am holding it up in the light, watching, measuring the level of activity. This is not a comb of quietly jostling bees; it's packed and it's busy … and there's something else puzzling about it …I can't quite work out what it is. Something odd. Suddenly I realise what it is: a needle-fine

'Queeping' sound. 'Eeeeeep ...pip ...pip ...pip.' Like morse code. Where is it coming from? It sounds again. A pulsar ... a radio beacon ... a message from space. In one corner of the frame a tight knot of bees are scurrying. Finally, I spot what it is they are covering: a queen cell, a dimpled wax bullet built on the bottom of the frame. The 'Queeping' is coming from this cell.

Here is a stroke of luck. A queen about to emerge, calling before she arrives. Some bee keepers call it 'Piping' or 'Quarking'. To my ear it's a 'Queep.' I lift out this frame and put it into a nuc box along with brood and food and some brushed-off nurse bees. I do this because this colony has a prolific egg-laying queen which I'm not yet ready to replace. So we start a new colony with the emerging queen. What is it she is saying when she sends out her little morse message? "Tell the lads to get airborne, I'm on my way"? Perhaps she's feeling the heat and needs air. Back home I tell Wife about the queen calling out from inside her waxy cell. "That's queepy," she says.

Evening

Late evening, actually. Dimpsy twilight. Momentary headlights twinkle from distant hills across the water. Out over the darkening sea night breezes are ruffling the surface. Across the bay a gold moon is climbing out of bands of cloud. This now is also a good time to be on the allotment, because there is nothing to be done for and about bees in the cool night air except watch the moon rise over the sea and listen to quietness. This leaves the mind clear and open. Without moments like this, moments of stillness and silence, life is a relentless and wearying repetition of habitual thinking. The mind needs space, needs a break from itself. It needs from time to time to be still. The most profound secrets of nature reveal themselves when there is stillness. Buddhists well-practiced in meditation understand this. But not only Buddhists. Today physicists and mystics have a meeting point which is at the centre of both their outlooks. Both of them see the physical world and every particle in it as one unified whole, not a collection of separate forms. All things interlinked, every manifested form emanating from the one source. Very much like a bee hive; a bee hive with a single mind. Somewhere along the clifftop a nightjar is purring. Ventilation for the mind.

10. Marmalade and the Meaning of Life

There was always something missing from the story of the feeding of the five thousand. I realised what it was recently. We have an excellent bread shop in the village, and on the slipway to the shingle beach there is a similarly excellent wet fish shop. (You're ahead of me at this point, but bear with.) So it's not uncommon some days to find I'm down the village main street garnering, yes, both loaves and fishes. Now bread and fish are all very well as good and worthy stomach fillers, but what did they do, the five thousand, when that craving for something sweet spread across their taste buds? I can buy myself a bounty bar, but what was there on the banks of Galilee to give a sugar-shot to the bloodstream? Where was he, the man with dessert in the desert, the beekeeper with his basket of dripping honeycombs, ready to have them blessed and divided up for some massed fingerlickin'? Imagine how that would that have gone down. Think of the applause. And what a recruitment gesture for bee keeping. But it didn't happen. That will always be the simplest two-course mega-feed in history, bread and fish, worthy, bland and good for the soul. A sound stomach filler in the search for the meaning of life. But missing a sweet.

For me, the search for the meaning of life is very largely satisfied in the discovery of a really good marmalade. Closely followed by maple syrup. Then honey. That sounds like a sacrilege in this magazine, but there it is. Each of those natural marvels has its own raft for carrying it to the mouth: porridge and pancakes for maple syrup; whole wheat toast for marmalade; scones for honey. You will disagree, and this is where bee keeping becomes engaging. And human. In bee keeping literature, the months April to September generate repetitive mantras of practical advice. But away from the mechanics of the hive, bee keeping isn't about insects, it's about human beings.

※

It's July and the population in the hives is now going down. This means that queen-substance which controls the swarming impulse is more effectively distributed among fewer bees. Nevertheless, swarm control inspections are still necessary. Bees can swarm as late as the end of September. Last year I laced my Autumn-feed syrup with a certain commonly available biostimulant. Composed of herbs, beet extract and molasses, the makers claim that it stimulates the colony, strengthens populations infected with nosema, has fungicidal and bacterial benefits and repels varroa. This is not a veiled advertisement, but I'd like to think that the supplement was partly responsible for the mass of healthy-looking bees I found on the Spring Inspection. Many bees, plus an embarrassment of over-wintered stores: supers filled with syrup, and late Summer honey, much of it granulated ivy. I lifted out several laden frames in March and replaced them with foundation, to make room for the Spring expansion. This made me think again about when to extract. This year I am adjusting to local conditions and waiting until later in August. Then feeding. Then varroa treatment. It's a balancing decision. Farmers have to do this all the time. Finding the warm and dry spell. Reasonably high summer temperatures are still needed to evaporate thymol crystals, which will make this particular anti-varroa treatment effective.

※

So here we are up early at the allotment, Wife and I. Wife has brought a picnic breakfast, her gardening gloves, her secateurs, and a magazine. Things look promising, as if I'm going to get some help, but it only looks that way for a few minutes. A few strategic snips of bramble stems, then she's settled into my chair with her magazine. This is my contemplation chair. To have someone else sit in it is more than an invasion of personal space. That chair, with me in it, is the hub, the think-tank, the policy unit, energy core of the multiple-branched operation that is my allotment. Usually it's just me here. I look out over the sea, I watch the coming and going of bees, I look some more at the sea, I nail together new frames, I think about things, I watch the sea, I don't think about things. Because that is the point of allotments, isn't it. Being alone. And the point of being alone is that you don't think about anything much. Obviously not everyone sees it this way.

"Don't sit too close," I warn her. "I'm taking out a few frames in this hive and they might just---"

She gives a small yelp and waves the magazine in the air. No, it turns out she hasn't been stung, she wants to read me something. "Here, listen to this. I've found just the present for you. When's your birthday?"

I tell her. Each year I have to tell her.

"It's perfect. This company offers a bespoke bee service. Honey bees at home. They come and do it all for you. They bring all the gear, the hive and everything. They fill it with bees. They do the inspections. They take off the honey and they fill the jars. All done. Sorted. You don't have to do a thing. How about that? Everything done for you. You just have to pay them."

I look at her. She can't see my expression behind the veil of my bee suit. "And that would free me up to do ... what, exactly?"

"Help me in the house, of course."

We break for coffee out of a flask, and rolls with bacon in, followed by rolls filled with the sort of marmalade that stops me searching for the meaning of life. Dark and bitter.

11. A Devon Summer in a Jar

January. Feet-up time for the beekeeper is coming to an end. The new book on bee keeping that you got for Christmas will have to be put aside. The bee equipment catalogues are interesting again. Things are looking up. Any sunny day now when you go and look at your bees it's possible you'll see some bright orange crocus pollen being trousered up the landing board. These are your stout little winter bees. They have endured a long life on low fat and high protein, with added constipation. These are the ancestors to thousands who are to pass this way throughout the long summer-to-come. There's a thought for any society bent on perpetual self-gratification. Every born worker bee flies and forages itself to death for the succour of populations yet to come, bringing home pollen-protein and nectar-sugars for the multitudes unborn. How's that for an example of the selfless life?

Notes from a Clifftop Apiary

Spotting early pollen in January may well be a phenomenon only experienced by us softies down here in the balmy southern lip of the country, where the flowering season is long and plants wake early. When do the beekeepers of Stromness see their first flowers? Here, my bees huddle in their winter group-hug within earshot of the wash of the sea, the tangled cries of gulls and in sight of a winter sun bouncing off the Channel waters. They're sensitised to the first crack in the winter cloud sheet, the first whiff of warmer salty airs. The great Wessex poet Thomas Hardy pondered on these things in his poem *The Year's Awakening*:

> How do you know, deep underground,
> Hid in your bed from sight and sound,
> Without a turn in temperature,
> With weather life can scarce endure,
> That light has won a fraction's strength,
> And day put on some moments' length,
> Whereof in merest rote will come,
> Weeks hence, mild airs that do not numb;
> O crocus root, how do you know,
> How do you know?

I am on my way through the village up to the allotment when Wife collars me outside the church. "Can we have a jar of your honey? We need some honey for a window display." Apparently there is to be a service with the theme of Nature in Our Village. Of course, I say. But I wonder aloud how long it's going to remain there, innocently on a ledge in the church window? "Isn't it likely to get pinched?"

Wife gives me a look. "This is church we're talking about, not some car boot table. Also, we want you to give us a talk."

"What about?"

"Bees. And honey. It's a service to do with Our Natural World."

This strikes me as a rather big subject. I give a cautious agreement. "Who am I talking to, and how long must it be?"

"It's the usual church congregation. About a minute."

It's my turn to give her a look. "You want me to say something interesting and important about bees in one minute?"

"Alright, a minute and a half."

On the morning of the service the pot of honey is still there in the window which now also has a nuc box on display, a frame of raised comb, and a bee suit dangling from a nail. There is something faintly disturbing about the bee suit hanging there.

I know when it's my turn to come up to the lectern because Wife gives me one of her steely-eyed looks from the choir stalls.

The congregation appears to be rather far away, concentrated in the six rear pews of the church. Why are they so far away? And doing everything except sitting still and looking attentive. They're fidgeting, glancing off to left and right, passing things to each other. It looks as if some of them are whispering jokes. Is this what it's like for the vicar every Sunday, holding forth to a distant group of people who are otherwise engaged in entertaining matters of their own making? And why aren't they sitting here at the front? I hold up a pot of honey and turn it in the sunlight from the windows. An English, no a Devon, summer – right here, distilled in a jar, I say. There are grins in the congregation from those people who regularly buy my honey. I say that a magical interaction has been going on between sunlight and flowers and insects and man to produce this jar of liquid gold. I mention insecticides and the loss of hedges and wild flower meadows. I say that our actions today have consequences tomorrow, and I leave them imagining a world where gardens are silent bee-less spaces, a possible future where supermarket shelves no longer hold honey jars.

"That was a bit sad," says Wife a little later. "I think they were rather shocked."

"Not at all," I say. "They were touched. The verger even asked me to repeat a part of it so he could write it down, because it was so moving."

"Which part?"

"When I said about how beekeepers are precious, and anyone lucky enough to live with or near one needs to really look after them and be very loving and affectionate."

"You better take that bee suit off the hook," she says. "It's scaring the children."

12. The Best Way to be Insufferable is …

It's 5.30 a.m. on a March morning and I am on the cliff. There is a cold white mist, but the air is filled with the cries of invisible seagulls and this is cheering. Below the fog the sea looks leaden. Half an orange moon is in the sky. Why am I here in the chill dawn? Half an hour ago I was warm and asleep in bed. But a thought woke me. *You left the strip of sponge blocking the hive entrance, didn't you?* Yes I did. Yesterday, shifting a hive from a rickety wooden stand onto some bricks, I blocked the entrance with a strip of sponge. All went well. But I forgot to pull out the seal. There was no going back to sleep after that. Not with the thought of countless pollen-laden bees locked out of house and home all night. Worse, of bees kept outside and frozen to death. I pulled clothes on over my pyjamas and drove through the sleeping village.

All looked well around the hive. No dead bodies lying around. No queue of petrified bees clinging to the box. Maybe they chewed-out a corner of sponge and squeezed back in? I pulled out the blockade, then stood back and spent a few minutes taking in the novelty of simply being there in the cold white dawn, under a gibbous moon. This, I thought, is not going to happen very often. I'm occasionally here on late summer evenings, when the light is at last failing and the bats are flickering about in the warm night air. But never at dawn. It seems to be a particularly self-aware time, being on the clifftop at the very beginning of the day, and at the very end.

That was a mistake, but mistakes are instructive. They mean much more than examples of perfect practice. Mistakes are human. To be vulnerable is to be human. Beginner bee keepers should be reassured about this. Kathryn Shulz's recently published book, *Being Wrong*, makes the point that mistakes are evolutionary, they open us up to growth. This is true of most skills in life. One Times Literary Supplement reviewer of this book put the matter succinctly: 'We take pride in being right. In fact, the best way to be insufferable is always to be explicitly right about bloody everything.' So how about this as a topic for one of the winter Branch Apiary Meetings: *Owning Up: Admissions of Failure. A Few Mistakes I Won't be Inflicting on My Bees Again.*

Another thing I won't be doing in a hurry is shifting a hive to a new but nearby site. Last Summer I spent a laborious and back-breaking time moving a hive one metre every two days. The new location was six metres away. The navigational ability of the honey bee is extremely precise. Bees will tolerate their house being moved a distance of a metre away from their old site in one go, but no more. Any further distance and foragers will drift to another hive, or will cluster at the original position of the hive entrance. The only thing that overrides the bee's acquired precision homing faculty is swarming. By absorbing the swarm impulse the bee wipes its homing programme clean, ready to receive a new imprint once a new home is occupied. This is why, when you catch a swarm from your own apiary and re-house it just a few feet from its old home, the bees accept the new site immediately.

The Times Literary Supplement isn't naturally a resource for bee literature, but occasionally one comes across items that catch the eye. A few months back the poet laureate Carol Ann Duffy published a marvellous poem in the paper, called *A Rare Bee:...out of the silence / I fancied I heard the bronze buzz of a bee.* And in January of this year I saw a piece by May Berenbaum, who is Head of the Department of Entomology at the University of Illinois, Urbana-Champaign. She was reviewing two new books: Peter Miller's *Smart Swarm*, (Collins) and Thomas D. Seeley's *Honeybee Democracy*, (Princeton University Press). At one point in her article, while discussing Miller's book, she makes the following statement: 'He probably doesn't realize that honeybees, held up in an earlier chapter as paragons of effective, orderly decision-making, also unsentimentally resort to cannibalism under poor foraging conditions, starting with workers eating the youngest larvae.' I questioned this in a letter to the TLS which was published in a subsequent issue, and so far Professor Berenbaum has not reasserted the observation that honeybee workers will eat larvae when there is little or no pollen and/or nectar to be had. It's possible she may have seen this behaviour under the intense scrutiny of academic research where artificially induced conditions may force bees to do this, but where else in bee literature does one comes across such a claim?

❦

My bees fly up and down the Jurassic Coast. I went to a talk recently by a man whose job it is to manage the coast line. How you actually manage the sea up against sections of clay, chalk, mud and shingle is a perennially incoming and outgoing challenge, but he has a strategy. He's put radio collars on some pebbles. He wants to track their movements up and down the 95-mile long stretch of cliff and beach. This opens up intriguing possibilities. We already tag birds and elephants and leopards. And prisoners on remand. And now we're tracking a pebble. How long before micro-technology produces a mite-sized camera attached to the back of a forager bee? New footage for Springwatch coming up. Our appetite for tagging the wild is limitless.

13. A Fine Looking Woman with a Beautiful Bottom

This column is called Notes from a Cliff Top, but actually, the cliff edge is several metres away, the other side of a more perilously sited domestic garden. But still, colleague-beekeepers hearing of a sea-edge location for the hives tend to be sceptical: "So, your bees have only got 180 degrees, half the compass to forage in, the rest is open sea. Bit of a handicap, that." As far as I can see, and on a good day I can see as far as Portland, it makes no difference. My bees aren't deprived, they fly inland to the field and lane hedgerows and to the rich pickings of village gardens, and they fly along the cliff edge contours where the hawthorn and the blackthorn and the bramble tangle with each other.

The hives are sited on the seaward side of the allotment under a blackthorn hedge for protection from prevailing westerly wind. This means that anything I try to do by way of vegetable-growing puts me in the flight-line of foragers. Digging or weeding, I'm an irritant, I'm in the way of flying bees. After five minutes I start getting head-butted. If that doesn't clear me off, then it's harrying, before finally a sting. Result: bees 1 : vegetables 0. I wonder if other allotment-beekeepers manage bees and vegetables together?

It's a sunny morning in mid-July and I am standing on a wobbly chair trying to lift off a honey super, but it's heavy and the chair is tilting. This is not a good arrangement; the hive stand is too high. At moments like this I have to admit that Wife's insistence that I put the mobile phone in my pocket is a good idea. I see myself lying winded in the grass with a brood box and maybe a super or two on top of me and just possibly a cracked rib; the mobile would come in handy then. This honey super is in fact a brood body; it's all that was available back in the end of May when this particular colony were roaring away, sliding nectar into the supers like a rising tide. Now I have a leaning tower and I have to risk my life to lift off the top and see if all that honey is capped and ready for extraction.

Cracking the seal under the top box and testing the weight, and keeping my balance, I sense I am being watched. Turning to look over my shoulder I see Wife standing silently by the scrub and bramble. The scrub and bramble is not the name of a pub. It is the state of my allotment. It is not usual for wife to be here. The allotment-apiary is a good ten minute walk from home. Why is she here? Why is she standing silently among the long grasses like a ghost. "Hello," I call. "What's up? Everything alright?"

"Man phoned," she says, keeping her distance. "He's got a swarm in his compost bin. Do you want it?"

"Did you take his number? I'll phone him back."

She is gone. And not a word about me teetering on a wobbly chair grappling with a heavy box full of honey. How is it that women never get the moment for concern and sympathy right? When you need it you don't get it; when it's the last thing you want, there's lashings of it.

In the evening I go to visit the man with the swarm in his compost bin. The bees are so docile I can scoop them out with my (gloved) hands and fill the bin lid, then tip them onto the sheet where they can walk into my waxy cardboard box. Later this is inserted into an empty brood box and a second brood body filled with frames is put on top. It's an alternative way of introducing bees to a hive; rather than knocking and shaking the bees down onto the top bars, the bees naturally move up and occupy the frames above them. I am indebted to Wally Shaw, who writes in Bee Craft, for this excellent practical way of introducing a swarm into a hive. Next day the cardboard box (which has been liberally painted with liquid wax) is removed and the colony fed syrup. Job done.

On one side of my plot is a cottage garden. The owner comes down from London at weekends. He tells me there is a bee hive installed on the roof of his office in Cavendish Square. I shall be interested to find out if he has a honey crop this summer. On the other side is Frank. Frank is my caustic vegetable growing neighbour. I ask him if he too is bothered by being in the way of flying bees.

"Nope. They's too high. They're over my head." He nods over at one of the hives. "Got a bit of newspaper stuck in there, you know that?"

I tell him I'm in the process of combining two colonies. "They're eating their way through the paper and making friends that way instead of fighting."

"So, what you give 'em then? The Telegraph?"

I shake my head. "I think bees are more likely Guardian readers, wouldn't you say?"

On the hive stand beside the Guardian-reading hives is a nuc, which I made up around a capped queen cell earlier in the summer. This caused me some puzzlement when, days later, I found a stunted little queen skittering over the cells. She was barely identifiable as a queen; it took me some moments of hard looking to decide that that was what she was. She clearly had the spider-like legs, but she barely had the abdomen. I marked her with a blue spot and continued to feed the colony. Later in the summer I cornered an expert, Dr Mick Street, in the Bee Tent at the Agricultural Show and asked him about runty-looking little queens and received the answer that what I was looking at was a newly emerged virgin queen, whose abdomen was not yet fully extended. Once she had been mated and her ovaries had developed, then she would start to look like a proper queen and the lateral membrane and the plate casings which make up the segments of her abdomen would extend and form the recognisable rear end. True enough. On looking in the nuc some weeks later I found her looking distinctly queen-like, gently bumping her finely-shaped rear end along over the cells. What could be more pleasing to the eye: a fine looking woman with a beautiful bottom.

There is another successfully expanding nuc in the apiary. This one was built up initially around a frame of eggs which had been lifted out of the parent colony and left for two hours in a brood box sitting on top of the hive. Time enough for the frame to become covered in nurse bees. This was then lifted out and put into the nuc box, with brood, food, emerging and shaken-in bees. They raised a queen from the frame of eggs.

Having hives overlooking the sea adds another dimension to bee keeping. Since much of one's time is spent peering into the minutia of the insect world, or else rearranging internal hive furniture, it's a relief to look up and out across the bay, to streaks of glittering light on the sea, and the travelling ridges of waves coming in out of the Atlantic, angled to the shore. And why shouldn't bees too share this pull of distance, even if they must stop short of the open gulf and harvest the very last and the very first of the flowers growing on the edge of this land.

14. *A Vision of the Afterlife*

The Irish writer Flann O 'Brien once made a vital observation. 'People,' he wrote, 'who spend most of their natural lives riding iron bicycles over the rocky roadsteads of this parish get their personalities mixed up with the personalities of their bicycle as a result of the interchanging of the atoms of each of them and you would be surprised at the number of people in these parts who are nearly half people and half bicycle...' I've had these words in mind lately after reading that the styrene molecule is now present in human fat. How did it get there? Quite possibly from styrofoam cups that keep your drink hot or cold, and from the plastic fluff-chips that fill the box in which your piece of delicate hardware travelled through the post. So by extension, what about polystyrene? What are the implications of this for the beekeeper who has opted for polystyrene hives? They're very popular in Finland, apparently. One can only wonder if after a lifetime involved with Nordic bees your Finnish beekeeper finishes up half-man half- hive?

March. The month of preparations. We can do regular inspections now to check for adequate food stores, to see if we have a queen laying normally, and to make sure there's enough comb space. Do we have the swarm-collecting gear ready? (Swarms can appear very early in the year down here in the balmy coastal margins of Devon.) Do we have enough supers for this year's incredible bumper harvest of honey to come? Are the Queen excluders sterilised and ready? Do we have spare boxes to house an artificial swarm? Do we have nuc boxes and frames ready for increase of colonies? Is this the year I replace the old bee suit? No. Not yet, I'm fond of the old propolis-stained, smoke-smelling, waxy jacket. I've had a few memorable experiences in that jacket.

My vegetable growing neighbour Frank is spreading seaweed on his allotment. Rather late, but better than not at all. His efforts look neat and professional: an even carpet of black ribbons strewn artfully all over his beds. This doesn't seem to work for me. Whenever I try it, bringing up a couple of sacks of storm-washed seaweed off the beach, I find I've got crab claws, scallop shells and fish heads in among the tangles. Clearly, I'm not sorting the stuff out properly. It's yet one more reminder that for me bee keeping and veg-growing don't work together. Not on the same plot. Obviously others in the country manage it, since there was an extensive survey of hives on allotments recently. Anyhow, for me I must focus on one thing. Or rather, several thousand things. Bees.

I'm looking at the hives that stand in direct sunlight and those in more-or-less permanent shade from hazel and blackthorn. The hives that catch the sun tend to do well. Is this because the bees are tempted out whenever sunshine hits the landing board? One might suppose that they have a better chance of remaining healthy because they can take cleansing flights more readily than their shaded neighbours. Plus there is less chance of fungus or mould. Photographs of Brother Adam's hives on Dartmoor show no trees or shade, only the boxes dotted about the open hillside, out in full sunshine. On the allotment, the sun swings around the sea's horizon and there is often a brilliant quality of water-reflected light. This is a motivation to cut back sections of the protecting blackthorn hedge and let some bright light onto each of the landing boards.

I had a near-death experience on the allotment not so long ago. My heavy stainless steel smoker toppled off the top shelf in the shed and hit me on the head. I was in there stacking supers prior to lighting sulphur strips to kill off wax moth. I suppose all the heavy moving must have tipped it over. Bang. An explosion of stars. The floor swung sideways and came up to lie against my face. The stars resolved themselves into a catherine wheel, and then into a tunnel. There was a light at the end of the tunnel and it was filled with hazy beings all clothed in white, and they were looking at me and beckoning. "Come," they said. "Come."

"No," I said. "I can't come. Not yet. My work here isn't done." I was beginning to find the white suits and the gauze in front of their faces rather familiar. "Who are you, anyway?"

"We are the Southern Chapter of the East Grinstead Bee Keepers Association."

"Good God," I said, "how come you're all here together, and why have you still got your beekeepers' suits on?"

"Ah," they said, "we were always a tight knit group, and we find the bee suits very comfortable, they sort of fit in here."

"Okay," I said, "but I can't join you. I'm in the East Devon Branch, and I don't see any of them here."

"Oh," they sighed, "the East Devons, they go on and on and on. They think they're in Heaven already---"

"---Anyway," I continued, "My work here is not finished. I think I need to spread some seaweed on my allotment. And besides, my wife wouldn't be very happy. I haven't shown her how to do an oil change on the car yet."

They looked at me thoughtfully. "We think you may be mistaken there, but we'll let it go. The choice is yours."

They faded away and were replaced by a single figure standing in the shed doorway with its arms folded. "What are you doing down there?" it said. "Having a sleep? Listen, let's get this oil change done, I've got shopping to do."

15. You Don't Want 500 Mars Bars, Do You…

Cold air and a grey sky up on the clifftop. Out at sea, a wintery sun finds holes in the cloud-ceiling, and shafts of sunlight slant down like theatre spotlights tracking over the choppy waters. Winter days on the cliff. Does an allotment 10 minutes' walk from home count as an out-apiary? It's a safe site. There's no livestock to knock the hives about, and this isn't woodpecker country, so the hives don't get disturbed. (Just how alarming would that be to a bee, I wonder, having a woodpecker hammering at the box inches from your ear?)

There are creatures here, rabbits, mice, gulls, pigeons, pheasants, a family of foxes, a rat or two, and they compete with all the gardeners on their plots for food. Fair enough. Everyone is aiming to get through the cold months intact. For the bees, winter is a reasonably easy time in the mild, temperate air of the south Devon coast. This is the margin of England where palm trees shake out their fronds in sub-tropical gardens. To the west of us a section of coast is called The English Riviera. Winters are rarely severe for us. Being up on the cliff means no frost pockets. And a fringe of trees shields the hives from the west winds. So all they need is sufficient food to see them through the cold weeks.

The greater part of those stores will be sugar syrup fed in September, after the Apiguard treatment; but some of it will also be late honey from ivy pollen and nectar gathered in October and November. Ivy is a mixed blessing. It's the final source of winter food. But a potentially lethal one. If it's a cold wet spring with limited flying weather, then a colony which is solely dependent on granulated ivy might starve. You could heft a hive in the Spring and find it satisfyingly heavy, but that's not a guarantee that all is well; the weight could be crystalline and difficult for the bees to dissolve and absorb.

If the presence of ivy stores is a perennial problem for you, then piling in the syrup in Autumn to limit the space left for ivy honey is one possible counter measure. There were many bright sunny days last October, and throughout the month I found noisy and excited bees crowding the landing board, jostling and swaggering up to the entrance with their baggy orange legs. This went on intermittently until late the following month. One note in my diary records: '24th Nov: yellow ivy pollen *still* being carried into hives.' Perhaps it's the peculiar quality of ivy honey plus the fanning of bees on warm and windless Autumn days that fills the air around the hives with the scent of honey. But how our weather veers. Within a week of that diary note, the whole of Britain went under frozen snow, bitter easterlies and sub-zero temperatures.

But let's not condemn late-season ivy pollen. Pollen is tough stuff. It's been around for over a hundred million years. The two thousand year old body of Lindow man recovered from a peat bog in Cheshire in 1984 still had mistletoe pollen in his stomach. That must have been quite a party. He'd not only drunk all the punch, he'd eaten the mistletoe too. And Christmas hadn't even been invented yet.

So here we are now on the edge of Spring. Yellow crocus and grey pendulous willow pollen will be appearing on the alighting board. The queen is shifting her laying rate up a gear. And over in the Branch Apiary the Beginners' Course is moving out of the cricket pavilion, where all the theory and the slides have been keeping them amused during the winter meetings, and is now gathering outdoors on Saturdays at the farm site. These have been learning experiences for everyone. Even for experienced beekeepers, meetings always contain hints, tips and suggestions for different ways of doing things. The prize for the most entertaining beginner's question goes to the lady who at the Q & A session asked: "Tell me, when bees go out foraging, do they always come back home in the evening, or do they ever stop off overnight at a B & B?"

Not so long ago the BBKA announced that from now on beekeepers could shop at Booker the wholesalers. This has been a very helpful move. At Booker a 25 Kg bag of white granulated sugar will cost you just under half the amount of money you would pay for the same quantity at the Co-Op. Most shopping trips contain an element of voyeurism, ogling attractive things you don't need and can't afford. A visit to the Booker warehouse is the reverse: you do need and can afford the big bag of sugar, but all the other stuff you trundle past with your reinforced heavy-duty catering for-the-five-thousand trolley is of no use to you. Cooking oils in tonnages; teabags by the five-thousand-in-a-box; rice by the impossible-to-lift sack. You can't buy a mars bar in Booker. You have to buy a box of five hundred. And you don't want five hundred mars bars, do you? Steady on, – No, you don't.

So I'm lugging the sugar sack into the kitchen, when Wife appears. "That new bee keeper down the road rang while you were out," she says. "Why are you bringing all that sugar into the kitchen, it'll only bring ants. Anyway, he wanted to know what hives and frames you use. I told him you had Longstrong hives and Hoffley frames."

I ease my aching back. "I see. You didn't consult the catalogue, then?"

"Don't need to, do I? I know about bee keeping."

"They're called Langstroth hives and Hoffman frames."

"Near enough." She eyes the sack. "I hope you'll not be dripping sticky syrup all over my kitchen floor. I don't know why you don't use baker's fondump."

16. Manuel of Beaky Ping

It's a windless glittery April morning and I am burning piles of cut brambles on the allotment. All around me a fine white ash is drifting silently to earth. To the east a thick mist has filled the valley all the way down to where the Axe river empties into the sea. My mobile chirrups. I keep it on chirrup so it doesn't clash with the chiff-chaffs, finches, robins and wrens who all live up here. A text from Home: *This book u wantd for yr birthdy. Can't read yr writing. Manuel? Who he? Wots Beaky Ping? Place in China? Can't make it out. Am off to yoga. Left you a crap sandwich in fridge.* Sometimes people, it seems to me, wilfully misunderstand you. I look out to sea and am reminded once again how uncomplicated life is on an allotment. The only sounds here are birds, and distant waves scrolling on shingle. A line of mist from the river mouth curves a mile or more out onto the shining sea, it lies there like a fat strand of plucked cotton wool. Since we live in a fishing village I hope that the sandwich turns out to have crab in it, rather than anything else. Back home I find it does. While I am munching it I read Wife's note about being totally unable to locate a book about a Manuel of Beaky Ping. I write underneath it: *You may have more luck if you try Manual of Bee Keeping*. Words can be such an imprecise way of communicating. I once heard a customer in WH Smith asking for a Pedge-Eddie diary. It took several goes before the assistant twigged that what was wanted was a Page-a-Day Diary. And I pitied the visiting English student in New York who spent a whole day in a city library trying to find the meaning of the word Narjus. Her flatmate was feeling Narjus. Turned out the word was that feeling when you are sick and feel like vomiting. There were in my teaching days some memorable misconstructions. Lucy, aged 13, once informed me that men who liked to wear women's clothing were usually called Transtightvests. And Daniel, 14, told me about a hitherto unknown work by Charles Dickens. "Yeah Dickens, yeah. I know about him. He's cool. Guy wrote Great Explaystations."

The writers of the Government Consultation Document on Pollination, which may have landed on your computer recently, also have a problem with communication, although they probably don't know it. Look at this, for example:
 ...over the next five years to 2019 {we intend to focus} *on improving understanding of the baselines for status and trends in pollinator populations and pollination services in England, and addressing policy-relevant gaps in understanding of pollinator interactions with crop production and wild flowers ...*

This is committee room language; public document language. People get to write like this when they're a long way away from the real things they're talking about. The writers should read George Orwell; he nailed the sloppy fashion of bolting ready-made phrases together to construct vague inflated ideas. A little further on, the document proposes a strategy (where would officialese be without 'strategy'?) '*to further develop the concept of payment for ecosystem services where gaps exist in developing this concept to apply to pollination services.*' What does that mean? And what are Pollination services? Planting pollinating insect-friendly flowers? Is my allotment an ecosystem service? Is the proposal here to pay people to keep bees? In that case, yes indeed, go ahead and pay us.

Very high in the General Public's list of loveable insect pollinators is the bumble bee. The Xerces Society for Invertebrate Conservation recently teamed up with Endangered Species Chocolate, a company that sources ethical cacao and makes chocolate with no guilt – at least not in the environmental sense. The point of the partnership between these two bodies was to set a competition for an endangered species to feature on a chocolate bar wrapper. So guess who beat the orang-utan baby, the panda baby and the wallaby baby? Bombus, of course. People just love that ball of fuzz that floats between your garden flowers. Gardens are fairly short on funny sights, but Bombus disappearing up a foxglove trumpet and then giving it a vibration burst to release pollen has got to bring a smile to your face. In The Hobbit, Tolkein comes close to naming one of the dwarves after the bumble bee, he is Bombur. Bombur is fat and clumsy and he falls asleep at crucial moments. But he is loveable. Bumble bees are not strictly the subject creatures of this magazine, but they do occupy the same evolutionary space in the scheme of life as our honey bees, -- along of course with all the rest of the pollinators, the uncountable populations of butterflies, lacewings, moths, flies and wasps.

The visit to our village of the Polish Youth Volunteers was coming to an end. Accompanying me for a day's work with the bees on the allotment had been Ludmila, a rosy-cheeked girl well-built for hefting hive boxes. We gathered at the coach pick-up point to say goodbye. Wife was hastily looking up words in a Polish phrasebook.

"What are you doing?"
"Finding something for you to say. In Polish."
She handed me a note with incredible words on it, and no vowels. "I can't say this."
"Yes you can. Make an effort. They like it when you try."
"What does it say?"
"It says Thank you Ludmila, You were the wind beneath my feathery wings."

I stumble through the line. There is a silence and I look up to find all the students gathered at the coach door are looking from me to Ludmila. The accompanying teacher sidles up to me, her eyes on the ground.

"Sir, you have just thanked Ludmila for breaking wind under your duvet."

17. Hearing Impairments and A Big Moth

Wife rings me to say she is standing in a bean queue. A bean queue? Are we into food shortages now? This is news to me. I didn't know there was a run on beans. "What? – Where -- Why are you waiting in a bean queue?" There is a stony silence. She says slowly and patiently. "I am not standing in a bean queue. I AM STANDING IN B AND Q. What was it you wanted me to get? I've forgotten." I had forgotten too. This is the stage we are at in life now. Mishearing and forgetting. Moving on… It appears that following the notice that I wasn't going to write any further pieces for this magazine, suddenly the editor became completely inundated with a letter from a lady in the Orkneys. She wanted to go on reading Clifftop Notes, she said, because she thought we had something in common, she and I: we both suffer from wind. She at the top end, me at the bottom. Of the country, that is. She pins her beehives to the ground with straps and tent pegs, and grows hedges and tree brakes around them. She likes to picture how it is with me on the South Coast, looking out over the Channel. Am I ever going to come up with a new solution for bees in wind? Well, sorry, but no. I have a black thorn hedge. And the tilt of the land is eastward, ie away from the prevailing wind.

It's a good place to keep bees, here on the soft clays and chalks and sandstones along the fringe of that great bite out of the underside of England, currently called the Jurassic Coast. The cliff-edge flora is immensely varied. The honey is multifloral. The farmers tend not to plant the sort of crops that need to be sprayed. And generally, we're in balmy land. Balmy airs on early spring days. Balmy sun on the sea. The tens of thousands of beekeepers in between Kirkwall and Lyme Bay may at this point be thinking it's all very well to batter on about beekeeping within sound of the waves that lap our shores at the top and the bottom of this country, but what about the rest of us? The truth is, there is no mean average experience of beekeeping practice in Britain. Whether you keep a hive on a barge that goes up and down on the tide, or on the roof of Debenhams (there are several on a flat roof in the centre of Exeter) or in the middle of a wood or in a suburban back garden, in essence, we all do seasonally exactly the same things. We keep our livestock healthy and productive. The same as does any food-producing farmer responsible for his animals.

My visit to the Orkney beekeepers is yet to come. But we do occasionally get some weather that excites the blood. "Storm force eleven, gusting to hurricane force twelve." It's not often you hear that on the BBC shipping forecast. I heard it on the night of 27th October. In the event, the main force of the blast drove through to the north of us. After it had passed I went up to the allotment to check on the hives. The sun was warm on the cheek, the innocent flat sea sparkling. I thought: here is a distinct plus; in this secluded spot at this height above the sea the bees get a double dose of whatever is present in heat and sunlight --- once out of the sky, and then again reflected off the sea. The hives were fine. But there had been an intruder. An immigrant from across the channel had landed right here on my allotment and had tried to get into one of the hives. Without success. The visitor had got stuck in the entrance and was firmly wedged in there. Dead. I pulled it out by the legs. It was an enormous moth with beautiful velvet wings coloured like pot pourri leaves, and a fat striped body as thick as my finger. It was *Acherontia atropos*, a Death's Head Hawk Moth. Migrants from Europe, they are quite rare in the British Isles, only some fifty or so sightings usually being recorded annually. Unlike most moths who are content to do the hover work and suck nectar up the proboscis from flowers, these have a reputation for invading bee hives in search of ready capped honey, piercing the cell with a short tough proboscis and getting straight on the money. This one, sadly, died with its head fixed in a bee-spaced gap. The Recorder for Butterfly & Moth Conservation in Devon reports that this specimen is the only confirmed sighting in the county of Devon in 2013. 2013 was a warm Autumn. The grass kept on growing. We were into late November, and local farmers were taking off a third cut of grass for silage. I was looking at the shaggy lengths of it, hacking it away from the hives and thinking about strimming when my mobile went off. Home calling. Wife: "On your way back will you stop in the village and get me a Balti?"

"A What? A Balti? A hot curry? From the Indian takeway? Are you sure? That's not like you. Have you got a sudden craving, or something?"

Another stony silence.

"What are you talking about? I said, would you get me a herbal tea."

18. On the Playing Fields of Bulawayo

"Is your bee keeping high end?" asks Wife over the breakfast table. She has been reading the weekend colour supplements.

"No," I say. "What does that mean, anyway?"

"It means you're Out There. You're on trend. You see, you should write one of those columns about a typical day in your life. Like these people in the back page of the magazine. It would be a way of publicising your honey. Like, you start the day with organic fair trade fresh roasted blue mountain coffee. Then you---"

"But I have a mug of builder's tea."

"---Then you move on to your own honey, mixed in with – no folded in with Greek style yoghurt, layered with sliced kiwano and papaya and sprinkled with goji berries."

"Is that a high end breakfast?"

She says she would find it totes delishballs.

It's distressing to break into and dismantle a thriving colony of wild bees. The photograph here shows beautiful lengths of comb over a metre long, tawny yellow, white and gold comb, and dripping balls of bees. Thousands and thousands of bees. They had been living for years in the cavity wall of a wooden chalet and we had to prise off plank after plank with a crowbar to expose the full length of the comb. I didn't like sweeping the bees with a soft brush into a skep; I didn't like seeing sagging slabs of comb break up and topple away; the walls of the city tumbling. The chalet was buried in undergrowth halfway down a wildly overgrown cliff overlooking the sea; it was a wonderful place for bees to live. But the owner of the chalet was intent on taking down the structure, burning it and rebuilding a modern version. Which is why I had been called in. The best option in a situation like that was to take away as many bees as possible and give them a home in a different location. I spent a long time up the stepladder studying clots of bees, looking for the queen. In the end I had to sweep them. I wired a portion of brood-bearing comb into a frame and that went into an empty hive on the allotment along with the loose bees. That was eight weeks ago. I've yet to spot a queen in that hive, even though there are eggs and larvae and capped brood. So either I was lucky and got the queen, or else the bees raised a new queen. Either way, I've grateful to have them; the genes of a thriving colony of wild bees is always going to be immensely valuable.

I was invited to give a talk to Probus. Probus is a club for retired old buffers who are at a stage in life where they all have pensions, politics and prostates in common. And to take their minds off these things they invite other old buffers to come and talk them about whatever particular buffery it is that the outsider know a bit about. My call-up was short notice because the Deputy Chief Constable of the County who was due to speak had been suddenly called away. Big shoes to fill, then. I'm sure I detected speculative looks of low expectation on the faces of the assembled Probertians as they eyed me sitting beside the Chairman's table. What would the DCC have talked about? Some juicy police cases. Looking at me, they must have thought what could this chap possibly have to say that would match the meaty experiences of the career policeman? I hadn't been invited as a beekeeper. Instead, I had been given a free hand. So I called my talk, 'The Other Half'. Some of the chaps, I learned later, took this to mean I was going to talk about my wife. That other half. No, what I had meant was the other half of life, ie some of the things I had got up to since retiring from teaching.

I was conscious that nearly all community gatherings from Rotarians to

Nursery School classes sooner or later get 'The Bee Talk.' For a start, I've never felt qualified to do that. And also, bee talks really require quite a lot of visual aids, some of them bulky. So I just touched on occasional moments when bees had injected some excitement into my life. Such as the time when I was umpiring cricket on the playing fields of Bulawayo. I had over the hours been noting how the shadow of the stumps got shorter and shorter until the moment when they disappeared entirely because the sun was plumb-overhead. I looked up – it was a moment between overs – and suddenly noticed that the only things standing vertical on the bright sunlit field were me and the stumps. Every player had hit the ground. Moments later I realised why: there was a growing hum in the air, a swarm of bees was crossing the open space at head height. I got down too. I wasn't keeping bees then, so never got to find out if African bees are noticeably more feisty than European bees.

Some of the remainder of that talk touched on the Rhodesian/Zimbabwean bush war of the 1970s, including an incident when my position on the edge of the Victoria Falls had come under machine gun and mortar fire from across the bridge on the Zambian side. Afterwards, over coffee, one blazered chap, slightly older than me, came up and told me he was one of the fellows on the Zambian side behind the shooting. His grin was noticeably broader than mine.

※

"Dunno how she puts up with it," say the ladies behind the tea counter at the Branch Meeting when I come to collect a cup of tea. "Your wife, she's a very nice lady, she don't deserve all those things you write about her."
I tell them I get her approval before I send anything off to be printed, but they are not mollified.
"That don't signify. She should have a right of reply: Notes from the Lady of the House. That's what's wanted. Put the record straight. She could tell a few things about what you're like to live with. What about that?"
I lead them to believe I'm thinking about the idea, otherwise I'm not going to get a cup of tea.
"Here," says one of the ladies, "I've got a joke for you. Ready? Time flies like an arrow. Right?"
"Okay," I say cautiously, "Time flies like an arrow."
"But fruit flies like a banana."
I give her a shrewd nod, and tell her it's a good one.

Notes from a Clifftop Apiary

15 Poems

Contents

Inspection

Swarm

Bottling Honey

In Late October

Two Hare Poems

The Dying and the Light

Great Pigs

Borderlands

Snipe

Winter Birdwatch

Even If

Spiders in the Garden

Road Tolls

Does

The Doe

Inspection

We lift away the roof to find seething on the comb,
a shouldering rustle of bodies in the mass, the scents
of heated nest, of ferments from a shuttered room.

A pulse of smoke scurries them down between rungs;
others take to air to harry us with their swing
and their sway, their strung, beaded warning songs.

We prise lugs free, ease up the loaded frames
into the sunlight, each glistens, drips,
thickened with honey cells down in the seams.

Under the trees the cruising swoop and skid,
as shadowed airspace populates
with circuits and feints about your head.

At my ear the pitch of fine needling,
barbed and livid in the electric air.
I wait, and wait for the burn of a sting.

What steers them around this bright charged space,
to weave this gauze about us as we work?
Some circuitry holds them to us, in aerial dance.

We move away, they clot our clothes, the walls of the hive.
Returning in the calm of dusk, a tap on their roof:
Anyone there?   A sibilant hiss warns of currents still live.

Swarm

Whose idea was this --- everyone off to the dance!
Some glee club, huzza-ing through the bright air,

They are sailing, cruising, crowding this noonday.
And what is they are singing? What are they writing

in their noughts and ellipses in the air, in their swirl,
cutting the peel off a round of space.

They drift across the blackthorn, a play of shadows,
being planet jazz, their fizz and skitter fading.

They leave us staring, shaken and dazed. Far off
their writing turns to scribble, a zig-zag skidding.

We follow: now they have clamped in silence.
A displaced people in flight grown wordless.

Space around them sags with the weight of the unspoken.
Their shadow among the leaves becomes terrible.

The thing that they have done has tightened on them.
Solemn now, weighted black, glistening and unexploded.

Horizons are pressing in. Their voltage is encased.
Dusk thickens. The stars are waiting.

Bottling Honey

Lugging off the deadweight of boxes
chock-full of honey from the hive
onto barrow into car into kitchen
leaves us broken-backed.

Here is the motherlode wrested away,
ballasted, as though we had to lift
this trove clear of the earth's pull,
the gravity-restrain of condensed nectar.

We tilt a half-bucket-full of honey
to feel the slow gather of its weight,
the sway of its heavy lava-crawl.

Later we jar it, slipping the valve to allow
a thick rope smooth as molten glass to come
sliding out, unravelling and then ravelling,
folding down into itself, filling jar upon jar.

We hold them up to the light to look
into a whole summer of sunlight held heavy in glass.

In Late October

In late October, under siftings of sunlight and leaf-shadow,
the murmurous come-and-go of bees around my head –
launching off and up and away into volumes of shining air,
to forage while I fan smoke and lift in turn the frames.

This black jostle of bodies, a people who live in darkness,
now lifted into sunlight; they stalk their cells, intent.
It's what they came here for. Every one a fuse-particle.
Every one burning. They smell of secrets, nested and fermenting.

One windless afternoon weeks back they filled the air
with their sailing, cruising, thin murmuring voices –
until some signal homed them in on a bramble bush, and they clotted.
A black ball, fizzing and dripping among leaves.

I blew smoke across the seeth of their fusion, lifted them,
laid them on a white sheet to coalesce with waifs and wanderers.
Later, tipped onto frames, the ball broke apart and sank away
like dark waters going to ground.

Now they return swagged and trousered with pollen sacs,
burrowing into the scrumble of bodies.

I am closing up their hive for winter. Seeing they have stores enough
to last them through. And still they fly, in the mesh of sunlight
and leaf-shadow, garnering and doodling in the evening stillness.
I leave them to winter in their golden darkness.

Two Hare Poems

I

One tawny stripe of weathered gold burns out there
among the feather and the foxtail grasses; he crouches
in the clover, folds back his ears and thinks himself well-hid.

Knows he is not – for see, he's off across the airfield
scooting, bumping on the oval gearwheel,
wearing the vault of the sky around his ears.

His world: cropped skylines, larks, and the glossy winds of heaven,
between the old runways, these dark lanes of silence
where heavy-bellied planes once taxied and took off.

And out of this stubble plain he draws other hares to him;
they limber up, run circles around each other, rag and chase,
tag and bob and canter in zig-zag stop-start lolloping.

What quirky characters! They pass around their single mind
in a hair's breadth: *You have it. No, you have it.*
Pivoting on a cornstalk.

From the edge of the runway I stand and watch them go,
their fitful limbs triggered to each other's whim --- until
one flees, speeding to leave his own body behind, outreaching

his own gallop. And they are gone in seconds. One after the other,
they have out-sped the joyous clench-and-spread of their own limbs,
and taken off below the skyline into emptiness.

In dreams I run like that. The bang and leap of blood in flight,
cutting through air like a blade, earth underfoot becoming space.
The way to leave.

II

The carefree hares are running along on their toes,
out across the level plains of the old airfield,
knocking up a fine dust, and we walk in their wake.

First they flatten in the stubble,
a fox-tint stripe of game fur, ears and supple limbs
compressing the earth under a dome of sky.

Then they're off, flicking away the spinning earth
beneath their toes.

They're running on wheels, while some whirring gyro
pins the eye to a point in space and the legs jink and traverse.

They stop. Cocked. In silhouette. Watching.

But the far hares can never be far enough.
They must run on and on, to lose themselves
between the transverse runways under the press of outer space.

The Dying and the Light

Coasting down the dirt farm track, tyres softly crackling grit,
I saw ahead a knotted fur-scuffle under the hedge,
something bowling itself about on plastered skins of wet leaves, twigs, mud.

And soon it was clear this was no idle skirl of wind-blown leaves.
Here were two bodies locked-in and hooked-up to each other
in a clamp of nails and claws and teeth.

Writhing in the mud beside the hedge, two bodies were working out
just how it would be that one of them would die.

I switched the motor off. Rolled gently to a stop. Sat.
A stoat spraddled across the spine of a rat, its fangs buried and burrowing,
working away deep in the neck of the rat.

They bounced and tumbled, they heaved and rolled,
a thin screeching in the air. The rat, clenched into a tight meatsquash,
rigid against the thing clamped around it, bounce-jerked at this spread
of fangs and hooks, hardened and burrowing tight into itself against this
excavating in its neck for veins, spine, nerves, anything to snap or bite through.

Their bodies humped and heaved, then lay quiet for some nerveless seconds.
Breathing. Then the struggle was on again, a keen squinnying electric nerve-
squeal staining the air like acid.

And the deeper deeper biting, the tighter bunched-up clenching,
a welded wrestling of these bodies, working as hard as ever they did at anything
in their lives at this death.

I had miles to go and so pressed on, leaving them: two kinds of burrowing.
I thought of meeting your death in a meeting of teeth.

I hope numbness quickly came,
that nerve-seeking needle teeth found and quickly cut the wires.

I recall it now while a spot of sunlight on my wall shifts in and out of focus, pales and wanes away to nought on heedless stone.
I hope that when the moment comes for me I'll see it clear for what it is ---
and go like that, not clenched against the points of darkness sinking into me, but go, like sunlight fading on a wall.

Great Pigs

The great pigs are jostling in liquid swill at the long trough.
Big nakednesses rubbing bristly leathers in the ruck.
Friction is crackling in the air, nerves are raw out here in the high windy acres
of the upland farm.

They let slip blurts from throats that grind soil and muddy stones.
They erupt to fine and lengthy outrage, rending the ammoniac air.
What else can they do here but lie slabbed-up together in a pile, sleeping in late
winter sun, waiting it out, brewing their bodies, leaning joints.

The visiting crows and gulls alight to inspect their turnover, then take to the
skies in droves; but the wired-in pigs must wade on across the hillslope paddocks,
lumbering hammy buttocks, pressing their tons pointedly into the mud, heaving
choice meats across a chewed-over waste.

The great pigs are fractious, loaded up with chemical kilos, they wear
their bodies like blisters; they are our fatted kine, bulked out on the rutty earth,
for the full English.

Borderlands

Amber cloud-light spills out from under a late December afternoon.
A freezing sun slides down behind the blackened tree-networks,
and over field-acres stolid pigs are ambling perimeter wires.

Above them squadrons of birds peel up and away from the mud wastes,
they rise and fall in the cold dusk airs; they light onto the suck
of slop-wallows, iced shitpools where pigs trundle their bow waves.

The pigs rub wads of gristle against each other, their punchbag bums,
their rows of baggy teats, earflaps, horny noses, -- through the swill ponds,
they lever up stones and set awash gouts of silt-creamed waters.

The failing light is filled with crows and gulls, a whirl of starlings;
along the field hedge stark oaks are studded with rooks in the branches.
Where is salvation for these creatures under the stain of spent light,

under a draining sky, the comfortless onset of winter's night?
The land is peeled; birds cling to the stalks of trees; pigs wade the wasteland.
Here is Outside. Regions where a thought dare not go. Borderlands.

Snipe

A numbskulling wind drives out of the north across the salt marsh.
We are treading the sponge matting of dead reed margins,
and the stands of dry sheaves sway and hiss about our heads.

Here now suddenly firing off in a silent flicker, snipe,
karooming away upward, scissoring the air, flick-flacking,
snicking off corners of space with their blades, to slice down

out of sight into some hidden side-channel.
Who would feel a lift of spirit like that out here in the reed beds,
between widths of wind-chipped waters in a February dusk?

We tread the rim of the mere, between faint blues of iced light.
Last night's freeze lies, a shatter of glass in the stalk matting.
In the wind, in the dusk, most things do not bear thinking about.

We tramp away in grainy dusk over sodden mosses.
Ahead of us the snipe still jink and pivot on their knife-points,
like flicker lightnings, like things remembered, and lost again.

Winter Birdwatch

Lapwings flick-flack, batting the air with their paddle wings;
teal flit and scurry, balance on platforms of iced waters.

A flock of godwits wheels, banks, turns to a flicker-net of lights;
curlew inspect the tide's leavings, pacing between mudstools.

Redshank's flute whittles away at echoes across the water;
snipe cut and snip into wind, sideslip to incise the salted marsh.

Tide sucks out the numb vein of the estuary, exposes velvet silts,
the creak and tick of mud channels, tussocks, a tacky mudfloor.

World comes unstuck from the riverbed, ice fields steaming in sunlight,
leaving a ripple of tideline where flotillas of gulls ride ruffled waters.

Across the valley hoar frost has blanched the brushwood trees.
It stays all day, silvered and pinking in the late afternoon.

By dusk the salt marsh ice pools are tightening in the reed beds.
Spars rebuild, braced for night, channels lock down under plated ice.

A dark tide refills the estuary. Across the saltings white silence
waits for the black breath of outer space.

Even If

In the skies above the upland mud wastes
where disconsolate pigs patrolling the wire
must come unstuck at each step in the quag,

and spend their days sucking mud off flints,
or else slumped together in a pile of slubber ---
in the skies above all this a single crow is flung

somersaulting in the wind. How should we fare
in this state? Beside the sunken lane, near a corner
of the wood, a stark oak is filled with crows all facing

into wind, like black ticks of approval at how things
have turned out. Squadrons of gulls are wheeling
over no man's land, steering their blades, dividing the wind.

Would we come unstuck? O but what if you and I
were to don boots and rain suits to cross these miry wastes,
and were caught in a rainstorm, and if you got boot-gripped

in the sludgy slurry pool, and I pulled and we both fell over.
That would be a thing to carry with me on through life,
even if you had never come around to love me.

Spiders in the Garden

I

It is September and spiders have the garden all sewn up.
Down in among the winding shrub-canyons
networks have been strung across the flight paths.

The sprung gauze wavers and bounces in the breeze.
Elastic, tensile, aerial ropeways, laddered airspace;
each has a bulbed spider in the bull's eye, crouching.

Their finickety nibblings are engaged in tying up parcels,
tasting soups; then it's lateral trampolining in the wind,
while blue lights slide like needles up their threads.

I am drawn again and again to go and squat and peer
at these mottled sacs, clinging buffeted to their tilting
galaxies – until, that is, moonrise, and a first star.

II

Spiders in the garden are making up time, ticking their way
around clock faces, joining taut lines between plants,
spreading networks that glint and belly in the breeze.

Their pinning of cords to a point in space, revolving tethers
to dab each joint with glue, swinging around the loaded pod
of their own gut frees them to spin out days clinging at the core.

Watching the shiver and ripple of lights on threads, the stillness
of the waiting thing, bulbed there, artist poised in the heart
of her own work, I stare through these slanted plates

of webbed light into lengths of unmarked time. Outside: winds
flex and nuances of light reverberate in the bounce and quiver
of spiders clinging. Inside: the building tick of the clock.

Road Tolls

We thought we were innocent, simply talking
as we drove along the country lane,
until a speckled something darted out
under our wheels. There was the bump,
the tumble, a flung parcel, and the mirror
filled with a whirling storm of feathers.

So then we stared a little longer
at earlier deaths along the route.
This rag of fur, bloodied, its content
of red seeds scattered like a burst fig;
that old pelt melting into a grass verge;
a knot of plumage pinned to tarmac.

We watched crows lifting at our approach,
up and away from raw meals, pick-pockets
at some slit skin purse, the owner long gone
under the hill. Something in the accretion
of these mute displayed bodies suggested
a price yet to be paid, a debt building.

Roads take a toll. Spillages of country viscera.
And now we raise our sights to cellophaned flowers
tied to posts, and we speak of that certain field gate
in Somerset where a constant blaze of blooms,
a floral rainbow, arches over a nest of sprouting vases,
bouquets in buckets, an altar of petals for a loved one.

Oh, and such a loved one, who died on impact there.
What to do with the love? The hole left
in life by death lets in cold, an ache of space.
We plug it with flowers, frail petals.
Alas, no flowers for the tumbled pheasant,
the bumped-off rabbit. Though, we spare a thought.

Does

I surprised her once in early Spring, her many faces at a chain link fence
grouped in silence, staring at me standing below in the sunken lane,

and the fathoms of all her eyes deepened, and a whisper flew
like wind through her bodies, a twitch, and she was gone.

A cloud of soft fur ruffled, she tipped her numbers away
across the grass to where they gathered in a corner, poised, staring back,
the flex of her many delicate forelegs held taut as a bow.

She is being bred here in secret for her fine, her dark, her inmost meat.

She sheds her sleek flesh, her coat, leaves them to us here as discards,

while elsewhere she slips between shade and shadow, printing damp mud with a
stiletto toe, laying down her lovely form in the wild garlic of the bluebell wood,

leaving a drumbeat echo of her flight under the calls of birds, under cathedral
trees.

The Doe

A windless morning of sunglitter on the estuary.
The mud creaks and ticks, and the waterline fills without us noticing,
a mirror lifting in silence.

Wigeon and godwit skewer the marsh grasses; curlew whistling
across the mudflats. This December sun is an echo of warmth
flashed off slow glossed waters.

Mid-afternoon, and we head home. Every fleck of this late winter sky
streams down to a knot, a tawny glow across ploughed fields,
ancient fires below the rim.

We round the bend in the lane to see lying stunned between hedges, breathing,
whited-out in the silent scream of newly fractured bones, the doe.

We stand in the dusk to see what is to be done.
Some voltage jolts her to panic scrabbling at tarmac,
the terrible stone road slamming her every way she tries,
falling into her face, grinding off fur, skin, blood.
She rows herself in circles, frantic to be fields away from us.
And the road goes on scraping her face to a bloodmess.

Covered at last, we lift her into the hatchback.
She batters at cabin dark, and the agonies of snapped bones.

Later, we pull up the hatch to find her curled like a cat
over folded legs. She is dying quietly in her own time.
She must have reached a place, for she lifts a slow head
and rests it in our palm, allows us to position her skull just so.

I recalled this in the night, in the early hours, when I looked out
onto an ice-crusted car, glinting in starlight
in a vicious silence of crystal air, and wondered how
she would have fared – under a blanket this night
in the boot of an ice-metal car, or in a wood lying with a mate
in a bole of leaf litter.

For now, though, we have interrupted her own dying
in order to kill her with a bolt through the head.
It leaves cordite in the air. And the sense of the fall and dwindle away
from here – which is the dark and ice and silence
of a winter night – of this doe.

'Looks like you had a dead body there,' wags a passing neighbour
in the night, as I sluice blood from car and tarp.
'Funny you should say that,' I say. (But not funny.)

I am waiting for some way to mark her moment; to see her free and intact,
and somehow already included in the shatter of sunlight on estuary waters,
in burial light across the fields.

A Note on the Poems

Some of the poems reprinted here have appeared elsewhere. *Inspection* was a prize winner in the West country Arts Council poetry competition in 2010, *Swarm* was published in Devon's Beekeeping magazine in July 2011, the first *Hare* poem was a prize winner in the BBC Wildlife Magazine poetry competition in October 2006, *The Dying and The Light* won first prize in the Bridport competition in 2001 and *Snipe* won second prize in the Bridport Competition in 2010, *Even If* was a prize winner in the Mirehouse Poetry competition in 2014, and *The Doe* was a prize winner in the Torriano Poetry competition in 2012. Other poems were published in *Poems*, a joint volume with John Torrance, published by Hooken Press in 2011.